Animal Math

Adding with Apes

By Adele James

Gareth Stevens
Publishing

Please visit our website, www.garethstevens.com. For a free color catalog of all our high-quality books, call toll free 1-800-542-2595 or fax 1-877-542-2596.

Library of Congress Cataloging-in-Publication Data

James, Adele.
 Adding with apes / Adele James.
     p. cm. — (Animal math)
 Includes index.
 ISBN 978-1-4339-5656-0 (pbk.)
 ISBN 978-1-4339-5657-7 (6-pack)
 ISBN 978-1-4339-5654-6 (lib. bdg.)
 1. Addition—Juvenile literature. 2. Apes—Juvenile literature. I. Title.
 QA115.J346 2011
 513.2'11—dc22

                                        2010046734

First Edition

Published in 2012 by
Gareth Stevens Publishing
111 East 14th Street, Suite 349
New York, NY 10003

Copyright © 2012 Gareth Stevens Publishing

Designer: Haley W. Harasymiw
Editor: Therese M. Shea

Photo credits: Cover, p. 1 Eric Cevaert/iStockphoto; pp. 5, 7, 9, 11, 12, 14, 15, 16, 17, 18, 19, 21 Shutterstock.com; p. 13 Dorling Kindersley/Getty Images.

All rights reserved. No part of this book may be reproduced in any form without permission in writing from the publisher, except by a reviewer.

Printed in the United States of America

CPSIA compliance information: Batch #CS11GS: For further information contact Gareth Stevens, New York, New York at 1-800-542-2595.

Contents

Great Apes! 4

Gorillas 6

Chimpanzees 8

Orangutans 10

Bonobos 12

Gibbons 14

All About Apes 16

Glossary 22

Answer Key 22

For More Information 23

Index 24

Boldface words appear in the glossary.

Great Apes!

Apes live in Africa and Asia.

There are different kinds of apes.

Let's add with apes!

1 ape + 0 apes = 1 ape

Gorillas

Gorillas are the largest apes. They have black skin and black hair.

1 gorilla + 1 gorilla = 2 gorillas

Chimpanzees

Chimpanzees live in large groups. They are very smart.

```
  2 chimpanzees
+ 2 chimpanzees
_____
  4 chimpanzees
```

Orangutans

Orangutans have red hair. They have long arms and short legs.

```
  2 orangutans
+ 1 orangutan
─────────────
  3 orangutans
```

Bonobos

Bonobos look much like chimpanzees. However, their arms and legs are long and thin.

```
  1 bonobo
+ 3 bonobos
_____
  4 bonobos
```

13

Gibbons

Gibbons are apes, too. They are smaller than other apes.

```
  2 gibbons
+ 3 gibbons
───────────
  5 gibbons
```

All About Apes

Gibbons and orangutans like to be in trees most of the time.

```
  3 orangutans
+ 3 gibbons
  ───────────
  6 apes
```

Gorillas, chimpanzees, and bonobos use their **knuckles** when they walk.

```
  2 gorillas
  1 chimpanzee
+ 2 bonobos
-----------
  5 apes
```

19

There are not many apes left in the wild. We need to **protect** them and their homes.

Count the apes in the photo. Add 5. What is the total?

Glossary

knuckle: a body part that joins a finger to the hand

protect: to guard

Answer Key

page 20: 8

For More Information

Books

Ellis, Carol. *Apes.* New York, NY: Marshall Cavendish Benchmark, 2011.

Kalman, Bobbie. *Baby Apes.* New York, NY: Crabtree Publishing, 2008.

Websites

Apes
anthro.palomar.edu/primate/prim_7.htm
Read about the many kinds of apes and find out where they live.

Great Apes & Other Primates
nationalzoo.si.edu/Animals/Primates/
Read about the apes found at the National Zoo.

Publisher's note to educators and parents: Our editors have carefully reviewed these websites to ensure that they are suitable for students. Many websites change frequently, however, and we cannot guarantee that a site's future contents will continue to meet our high standards of quality and educational value. Be advised that students should be closely supervised whenever they access the Internet.

Index

Africa 4
Asia 4
bonobos 12, 18
chimpanzees 8, 12, 18
gibbons 14, 16
gorillas 6, 18
orangutans 10, 16